Did you enjoy this issue of BioCoder?

Sign up and we'll deliver future issues and news about the community for FREE.

http://oreilly.com/go/biocoder-news

BioCoder

FALL 2013

O'REILLY® Beijing · Cambridge · Farnham · Köln · Sebastopol · Tokyo

Contents

Welcome to BioCoder

Welcome to the readers, welcome to the editors, welcome to the contributors. We're all glad to have you along for this trip. A few words on what this is for, how it started, and perhaps where it's headed. Though we don't really know where it's headed. We'll find that out on the journey.

I've been following biology for a few years now. I'm not a biologist, and haven't taken biology since 7th or 8th grade, over 40 years ago. But I have watched it from a distance, and increasingly, it feels to me like something that's about to explode: it feels a lot like computing did in 1975, before there were PCs, but when a friend of mine got an 8008 and a used teletype and built a computer in his dorm room. That computing event signaled a lot of things. In '75, computing was arguably already 25 years old. But up to that point, it had been done by people with PhDs, people who wore white lab coats, people who had inconceivably large amounts of funding and built machines the size of houses. What my friend did was demonstrate that computing wasn't the property of a priesthood with lab coats: it was something that anyone could do.

We're now seeing that same shift in biology. Students are making glowing E. coli, both at community labs like Genspace (*http://genspace.org*) and BioCurious (*http://biocurious.org*) and in high schools. iGEM (*http://igem.org*) exposes thousands to the idea of building with standard biological parts. We're seeing community laboratories spring up all over the world: literally on every continent except Antarctica. Experiments that formerly required a fully equipped laboratory with hundreds of thousands of dollars worth of equipment can now be done in a community hackerspace, equipped with little more than a homemade PCR machine and centrifuge powered by a Dremel tool. We're discarding the lab coats (if not the purple rubber gloves).

One unfortunate theme has stood out in my many conversations with biologists, though: "Yeah, I think something is happening in Hoboken (or Austin, or Juneau, or Calgary), but I don't know what." While there is a large and active biohacking community, it's poorly connected. The players don't know who each other are and what's going on across the community. It's symptomatic that one of the world's largest gatherings of DIY biologists was organized by the FBI. They were friendly, but that's not the issue: if it takes the FBI to bring us together, we aren't talking to each other enough.

That's the problem we're fixing with BioCoder. BioCoder is a newsletter for synthetic biologists, DIY biologists, neuro biologists—anyone who's interested in what's happening outside the standard academic and industrial laboratories. We plan to include:

- Articles about interesting projects and experiments, such as the Glowing Plant
- Articles about tools, both those you buy and those you build
- Visits to DIYbio laboratories
- Profiles of key people in the community
- Announcements of events and other items of interest
- Safety pointers and tips about good laboratory practice
- Anything that's interesting or useful—you tell us!

Not all of the above, not in every issue. But as much as we can. We'd like to publish BioCoder quarterly. Although the first issues are free, we'd eventually like this to become self-sustaining.

To do that, of course, we'll need contributions. Send a note to me or Nina DiPrimio, our volunteer editor, and we'll set you up. If you're unsure whether your idea is good enough—it probably is, but feel free to pitch an idea before writing it up.

biocoder@oreilly.com

Thanks again. I'm thrilled to see our first issue, and I'm looking forward to the second.

—Mike Loukides
O'Reilly Media

Biotech's Cambrian Era

Ryan Bethencourt

As I write this article, I'm reflecting on the long expanses of otherworldly playa I've just left, watching sandstorms pass in front of me while in altered mental states and contemplating the future of our beloved biotech industry.

I have, until recently been living a double life with one foot in the corporate biotech world and another deeply in the world of biohacking/radical science (working on DIY biolabs (*http://www.nature.com/naturejobs/science/articles/10.1038/nj7459-509a*) and equipment, longevity research (*http://www.genescient.com/*), and ALS therapeutic development (*http://www.bizjournals.com/sanfrancisco/print-edition/2012/07/20/biohackers-go-solo-in-quest-to-find.html?page=all*)). I believe in the principles of citizen science and shared (or at least *leaky*) IP as a means of accelerating scientific progress, but I felt I needed to play my part in the "real" biotech industry. That changed three months ago when I realized that to create the innovation we want in biotech, we may have to burn the bridges that got us here and re-create it ourselves, with or without the dinosaur the current biotech industry has become.

Since 1978—arguably the birth of the biotech industry when Genentech created the first GMO producing insulin (*http://en.wiki pedia.org/wiki/Genentech*)—Biotech has become profitable and also heavily regulated. Biotech venture capitalists, the original sources of risk capital, have become risk-fearing middlemen/women who have been cowed into seeking safe returns for their masters (limited partners) and obsessed with spinning the right story to their customers (big pharma/biotech companies). Much of the shift away from risk has been rightfully laid on the FDA's door for an increase in regulatory burden and uncertainty

(http://www.forbes.com/sites/theapothecary/2013/08/31/how-congress-encourages-shortages-of-cancer-drugs/) that has spread as best practice globally and mired the pace of innovation. Regardless, large corporations and academia can no longer be entrusted to move radical science forward—their world has become a world of committees, budget allocation negotiations, and quarterly/yearly cycles, lacking in vision and with fear of failure. So where does it leave us? The refugees of the Biotech Valley of Death?

The power they've taken from the people will return to the people, whether these vested interests want it or not. Biotech and medicine have advanced at a glacial pace, but a massive disruption is coming that will destroy the antiquated business models in the biotech, monopolistic healthcare, and pharmaceutical industries. As technology's pace continues to quicken, the biotech industry is beginning to benefit from a digitization of biology, the maker movement, quantified self, grinders/transhumanists, crowdsourcing, and a resurgence in local production technologies like 3D printers. A small group of hobbyists (several thousand globally) has emerged over the last couple of years and has begun building biotech equipment for 1/10th to 1/1000th the cost, creating novel open source diagnostic/medical devices, and therapeutically experimenting on themselves, as well bootstrapping and forging new paths in bioscience, like creating commercially available, genetically modified, glowing plants.

The business models for these emerging biotech industries are still evolving, but a true hunger is emerging from consumers and patients for new products offered through crowdsourcing sites, such as microbiome analysis; cheap and effective hormone analysis; novel industrial enzymes; algae-powered lights; true disease modifying therapeutics for established diseases and therapeutic life extension; cheap DIY biolab equipment; and technologies that amplify human senses, like the electromagnetic implants *(http://blogs.discovermagazine.com/sciencenotfiction/2011/01/05/biohackers-and-grinders/#.UiW8bGRgaFQ)* pioneered by grinders (i.e., those willing to biohack their own bodies). PWC has estimated that the biotech industry will be worth about $1.2 trillion globally by 2020 *(https://emarket ing.pwc.com/ReAction/images/West_Coast_Tax_Roundtable_Presenta tion_10_26_11_FINAL_share.pdf)*, but this is based on a very conservative view of the industry, and with radical disruption and the creation of new products like synthetic meats, regenerative medicine, unconventional materials, and industrial enzymes, as well as the potential for homegrown biofermentation of many other products, the market for the "new" biotech industry will be vast and shifting.

In healthcare, I look forward to seeing the oligopolies that have stifled innovation and kept patients' healthcare prices high and access lower to come crashing down as our fellow biohackers create innovative ways to allow people to ferment their own products; even getting FDA approval for a novel drug will no longer be a practical issue if you personally have control over the means of production. As biohackers, we aren't interested in preserving the status quo but in overthrowing it for the betterment of humanity. The homebrew computer club was the past—the future belongs to those who homebrew biotech!

@ryanbethencourt (https://twitter.com/RyanBethencourt) is a biotech entrepreneur engaged in the war against the ravages of time and disease. He's working to accelerate innovations in medicine through biohacking, open innovation, and collaboration. His primary areas of expertise are human translational medicine, genetics, longevity, and biohacking. He can be reached at http://BamH1.com.

DIYbio and the "New FBI"

Michael Scroggins

In 1936, sociologist Robert Merton wrote an article titled "The Unanticipated Consequences of Purposeful Social Action." Merton argued that all purposeful action in the social world creates series of unanticipated consequences that trail the action as a wake trails a ship. Per Merton, this quality gives human action a reflexive character; consequences unforeseen at the initiation of a course of action often affect the very course of that action. Whereas some consequences are serendipitous, others are more ominous.

Which brings us to the 2012 FBI/DIYbio meeting in Walnut Creek, CA. During the three days of meetings, multiple FBI agents addressed the conference attendees and explained that they work for the "new FBI," not the "old FBI." At key points, FBI agents rose to their feet and gave the kind of personal testimony to the difference between the old FBI and new FBI that you might expect in an evangelical church. Whereas the old FBI was a policing agency that busted down doors, the new FBI is an intelligence organization that gathers, sorts, and most importantly, classifies information based on forecasted threats. The old FBI wanted to cuff you; the new FBI wants to get to know you.

The line demarcating the old from the new, per the agent's testimony, was the events leading up to the 9/11 attacks. It was not the actual attacks per se, but rather the activities of the attackers in the months leading up to 9/11 that sparked the change at the FBI. The hijackers' flight training was referenced at multiple points as an example of something that the new FBI would be aware of through relationships forged with flight training instructors and that the old FBI would, and did, pass over. Which is to say, if you are doing something with biology outside of established institutional boundaries, then you will probably meet your local FBI agent at some point.

A serendipitous consequence of the FBI's interest in DIYbio was the conference serving as a vehicle for the DIYbio community to meet face to face, which would be impossible without FBI sponsorship. While the FBI was busy lobbying attendees about the benefits of getting to know their local anti-terrorism agent, a counter education was taking place during informal get togethers outside the conference venue. A good portion of attendees must have taken note of the "FBI visit" instructions during a side trip to Noisebridge, a San Francisco hackerspace. Even more discussed strategies for working with or around the FBI in between more prosaic discussions about finding suitable landlords and insurance agents, talking to the media, dealing with local regulatory agencies, and attracting/vetting potential lab members.

A more ominous consequence was the implicit connection drawn by the FBI between DIYbio laboratories and flight schools. The new FBI assumes that DIYbio laboratories may be breeding terrorists along with bacteria. While the FBI is out to build friendly working relationships with DIYbio, the tacit admission that a DIYbio laboratory is a potential threat and amateur biologists, if not assumed guilty, are not assumed innocent either, is a new and consequential fact of living with the new FBI; the FBI directorate covering DIYbio falls under the rubric of weapons of mass destruction—and the harshest punishments the US government can offer. Finally, it is not at all clear how the FBI keeps tabs on the DIYbio community or with which other US government agencies (or foreign governments) they might share information. As Merton warned: "Here is the essential paradox of social action—the 'realization' of values may lead to their renunciation."

Michael Scroggins is a Ph.D. candidate at Teachers College, Columbia and a researcher at the Center for Everyday Education. He is currently in Silicon Valley conducting research for the project "Education into Technological Frontiers: Hackerspaces as Educational Institutions." He can be reached via email at michaeljscroggins@gmail.com or at http://about.me/michael_scroggins.

"Is this Legal?"

Patrik D'haeseleer

"Is this legal?" is probably the #1 question people ask us when they hear about the Glowing Plant project (well, after "can I have one?," of course). The short answer is yes. But the long answer is far more interesting.

Regulatory oversight in the US over genetically modified organisms (GMOs) is covered by an alphabet soup of laws and agencies. Different rules apply when you are dealing with a GMO food crop such as soy (covered by the Food and Drug Administration [FDA]), a microbe, an animal, or anything that has been engineered using a plant pathogen. I would hesitate to call this a patchwork quilt of regulations, because a patchwork quilt isn't supposed to have any holes in it, and these regulations definitely do: big, ragged, oddly shaped holes.

Most plant genetic engineering has historically been done by taking advantage of the plant pathogen Agrobacterium tumefaciens. You may have seen Agrobacterium at work if you've seen a tree with a large outgrowth on its trunk. Agrobacterium is a bacterium (duh) that infects plants. During infection, it injects some of its own DNA into the plant, subverting its host's machinery to make a nice little home for the bacteria. Plant genetic engineers have been using that trick to their own advantage by engineering Agrobacterium to inject whatever genes they want to insert into the plant instead. However, the US Department of Agriculture (USDA) has been understandably cautious about releasing any plant that has been infected by this pathogen, especially an engineered strain of Agrobacterium that could potentially infect other plants in the environment. In fact, this has historically been one of the primary justifications the USDA has used for regulating GMO plants.

Now, it turns out that Agrobacterium isn't the only tool genetic engineers have at their disposal to get genes into plants. In 1987, Klein and Sanford discovered that you can literally fire tiny bullets loaded with DNA into cells using an air gun. And when I say tiny, I do mean tiny: the usual ammunition for this "gene gun" is gold nanoparticles that are 1/100th the width of a human hair. Each gold particle is coated with strands of DNA coding for the genes we want inside the cell. The use of gold allows the bullets to be much smaller than the size of the cell, yet heavy enough to carry enough momentum to pierce the tough plant cell wall. This is the same reason that real bullets are made out of lead (or the even heavier depleted uranium) except that the gold particles will be inert once inside the cell.

The use of this gene-gun technology to circumvent the USDA's regulations on non-food plants did not escape the notice of the 800-pound gorilla in the field of plant engineering. Monsanto, in collaboration with Scotts Miracle-Gro, has been developing a bluegrass strain (the lawn variety, not the banjo variety) that was engineered to be resistant to their favorite herbicide, glyphosate (a.k.a. Roundup). Because nobody but your dog eats lawn grass, it's not covered by FDA regulations, and since they used gene-gun technology instead of our friend Agrobacterium, it's not covered by USDA's plant pathogen–based regulations. Scotts/Monsanto saw a huge gap in GMO regulations and waltzed right through it! Mind you, there were still plenty of voices saying that they should never have gotten away with this. After all, there are plenty of weed grasses that their bluegrass could potentially outcross with, and by inserting the herbicide resistance genes, they've given this grass an evolutionary advantage wherever there are traces of Roundup in the environment. But get away with it they did: the USDA ruled that their bluegrass did not pose a risk to become an agricultural pest, and that was that.

Now, compare that Roundup Ready bluegrass with our little Glowing Plant: Arabidopsis is not a very hardy plant, and since it is self-pollinating, it is highly unlikely to outcross with more vigorously growing weeds to begin with (unlike grasses). Also, rather than giving it a fitness advantage by making it resistant to herbicides, the genes we're inserting into its genome will drain a small amount of its energy to produce light, so it will likely do slightly worse than its unmodified cousins in the environment. Other than that (and the fact that Monsanto is a multibillion dollar company with thousands of lawyers), the two are fairly analogous.

So where Monsanto waltzed through the regulatory gap, we will be happy to sneak through after it and give you something you really want: not just another water- and herbicide-guzzling lawn, but a glowing garden of bioluminescent plants.

Mind you, I have nothing against rational, sensible regulation of genetically modified organisms. This is after all a very powerful technology. We also regulate car manufacturers, because we prefer our cars not to fall apart on the roadway. But if billion-dollar companies can get away with bringing an herbicide-resistant grass to the market without any regulatory oversight, then surely our ragtag band of biohackers should be allowed to create a little glowing plant as well?

So, that was the long answer: yes, what we're doing is legal. We have talked to the relevant regulatory agencies, and Monsanto already set the precedent with their

Roundup Ready grass. There is still a small possibility that our Glowing Plant project might get shut down by one of the alphabet soup agencies, but then they'd need to reverse their decision on Monsanto as well. And if a bunch of DIYbio amateurs are able to insert some more rational thought into the national debate around GMO regulation, then, personally, I wouldn't consider that a bad outcome either.

Patrik D'haeseleer is a bioinformatician by day, mad scientist by night. He is the community projects coordinator at BioCurious, cofounder of Counter Culture Labs, and scientific advisor of the Glowing Plant project, neither of which is in any way related or funded by his day job at the Lawrence Livermore National Laboratory and the Joint BioEnergy Institute. The views presented here are his own and do not represent those of the Glowing Plant project, BioCurious, CCL, LLNL, or JBEI.

DIYbio Around the World

Noah Most

The customs officer raised an eyebrow. I carefully explained the nature of my stay in Canada. My arrival in Victoria on July 26th represented the first stop on a year-long worldwide tour of do-it-yourself biology (DIYbio). I'll get my hands dirty in community biolabs as well as examine the social, ethical, and legal questions that are tied to the movement. It just may not be the easiest thing to explain when crossing a border.

I graduated last May from Grinnell College in Iowa where I studied biology, economics, and entrepreneurship. As I was exploring synthetic biology, I stumbled upon DIYbio and was immediately captivated; it seemed to fuse all my interests together beautifully. I applied for a Thomas J. Watson Fellowship, which allows for a year of independent study outside the United States. My proposal generated more odd looks—this time a good thing—and was just weird enough to be accepted. As a result, I'll be sharing my experiences as I visit different DIYbio groups around the globe, spanning from the United Kingdom to Indonesia.

Canada made for an obvious first stop, and Derek Jacoby, founding member of the Victoria Makerspace, welcomed me to my inaugural lab. Immediately, I was thrown into an introductory class where the students made E. coli express green fluorescent protein from a jellyfish species. As a summer fellow at Carnegie Mellon University, I had done this lab once before, except this time the other participants weren't undergraduates. Instead, I was joined by a woman with a background in communications, two high school students, a network specialist, and an onlooking real estate developer. Everyone was pleased to see their own colonies glow under a UV lamp. Genetic engineering by amateurs—it's not just something people envision in TED talks.

For an independent project, I am exploring DNA origami, which may have profound implications for DNA nanotechnology. This field was established in 2006 by Paul Rothemund, who figured out that a long, single-stranded DNA "scaffold" can be guided to fold with many different "staple" strands, which allows the generation of highly specific 2- and 3D nanostructures.[1] Essentially, if you can conceive of a shape, you can make it out of DNA, and all sorts of fascinating applications are being explored.

For example, Douglas, Bachelet, and Church built a nanorobot for delivery of molecular cargo.[2] Essentially like a clam shell, the inside features binding sites for cargo, such as an anti-cancer drug. The clam shell is locked with an aptamer that's bound to a partially complementary strand. When this aptamer encounters its target molecule, it preferentially binds to it, opening the lock. That target molecule, ingeniously, can be a protein found on the outside of cancerous cells. Thus, the clam shell releases its molecular payload upon just the bad guys that need to be killed. Such specificity may dramatically reduce the toxicity of the drugs. For my own project, I designed a proof-of-principle 2D light bulb, which Bachelet was kind enough to review. Next, I'll build a 3D design.

To what extent is DNA origami feasible for DIY biologists? There are questions as to whether DNA origami can be made cost accessible, but the method does have some advantages. First, with student status, the design software caDNAno, built on top of Autodesk's Maya 2012, is free. Second, the whole shebang is done in a one-pot reaction, so reagent costs are minimal. Third, the single-stranded scaffold is commonly the M13mp18 virus; three trillion copies can be purchased for about $30.[3]

The largest cost limitation to building the nanostructure are the oligonucleotide staples. Fortunately, in many cases, expensive oligonucleotide purification is unnecessary.[4] More typically, a set of staples is likely around $700 at the 25 nmol scale.

DNA origami requires access to an atomic force microscope for 2D imaging or a transmission electron microscope for 3D imaging. Microscope access may be the largest obstacle for any DIY effort. Fortunately, the Victoria Makerspace is working out access to these microscopes through an agreement with a local university, so it is conceivable that others can come to similar arrangements.

[1]. Rothemund, P. W. "Folding DNA to create nanoscale shapes and patterns." *Nature*, 440(7082) (2006): 297–302.

[2]. Douglas, Shawn M., Ido Bachelet, and George M. Church. "A logic-gated nanorobot for targeted transport of molecular payloads." *Science* 335.6070 (2012): 831–834.

[3]. Rothemund, P. W. "Design of DNA origami." In Proceedings of the 2005 IEEE/ACM International conference on Computer-aided design. IEEE Computer Society, May 2005: 471–478.

[4]. Rothemund, P. W. "Folding DNA to create nanoscale shapes and patterns." *Nature* 440(7082) (2006): 297–302.

It is my hope that this project will elucidate DNA origami's potential for DIYbio. Even if the wet work of DNA origami is too costly for many groups, the field is still idea- and design-accessible. Like with iGEM, undergraduates are already forming teams, producing designs, and executing projects through the BIOMOD competition (*http://biomod.net/*). I have compressed a crash course in DNA origami to 20 minutes, and it will be followed up by another lesson in caDNAno.

By the time of my next entry, I will have crossed the Atlantic and joined the Manchester Digital Laboratory in the United Kingdom. I'll report on a DNA barcoding project as well as my work exploring the social, ethical, and legal issues that surround DIYbio. As I travel, I'll begin to develop an understanding of the differences that exist between DIYbio groups and the environments in which they exist. How do DIYbio groups differ in their interests and projects? How do legal considerations change? How do local mores affect public perception of DIYbio and GMOs? I will share what I find as I continue my global voyage.

Noah Most graduated last May from Grinnell College where he studied biology, economics, and entrepreneurship. He has performed research in labs across the country and led a microfinance nonprofit that was hailed as a "Champion of Change" by President Obama. Through an interest in synthetic biology, he discovered DIYbio, and, six months later, he won the Thomas J. Watson Fellowship in order to study it around the world for a year.

Better, CRISPR Homemade Genomes

Derek Jacoby

The point of synthetic biology is editing genomes—adding and removing functionality until the organism you are creating behaves exactly as you want it to. Since this behavior is determined by DNA, these additions and deletions must take place in the genome.

When I was first getting into synthetic biology in 2008, some friends and I formed an iGEM team at the University of Victoria. Everything we worked on was based around BioBricks: biological units of defined functionality. BioBricks contain specific DNA sequences at each end of the DNA so that they can be precisely snipped out with restriction enzymes and then put back together in the order you want them. I remember being amazed at the specificity of restriction enzymes—to be able to cut a piece of DNA at a very precise point confers an amazing ability to move segments of DNA around as functional units and create genetic circuits. However, all the circuits we created had to be on small, circular segments of DNA called plasmids. This introduced some significant limitations: we could only create new circuits, existing functionality in the chromosomal DNA was inaccessible to us, the plasmids had to contain antibiotic resistances, and if we ever placed the new constructs in an environment without that antibiotic selection, our hard-won changes would be gone in just a few generations. So everything we created felt temporary. And it all came about because our palette of restriction enzymes was fairly limited. This is rapidly in the midst of changing.

Before talking about how that change is occurring, let's delve into restriction enzymes.[1] Evolved as a bacterial defense mechanism, a particular bacteria generates an enzyme that cuts a specific segment of DNA. Most frequently, it cuts DNA pos-

---

[1]. Pingoud, A. and A. Jeltsch. "Structure and function of type II restriction endonucleases." *Nucleic Acids Research* 29 (2001): 3705–3727.

sessed by an invading bacteriophage—a virus that infects a bacteria—so that the phage DNA is disrupted without harming the bacteria itself. Our menu of restriction enzymes is derived from these evolutionary battles between bacteria and phage and are named after the bacteria they are isolated from. For instance, EcoRI (pronounced "Eco-R-One") is the first restriction enzyme isolated from E. coli and cuts precisely at the sequence GAATTC. About 600 of these enzymes are available commercially, which provides a significant set of options for how one can design and manipulate synthetic DNA. But it does no good when trying to cut a segment of DNA that does not match up to one of the available enzymes, so it is of very limited utility in editing natural chromosomal DNA.

Restriction enzymes are DNA-binding proteins, and the activity of a protein depends largely on its shape. Although protein structural prediction is improving, it is still a very difficult problem, so creation of artificial restriction enzymes to cut at a particular point is nontrivial. Some approaches to creating sequence-specific DNA binding proteins do exist, though. The first to arrive on the scene were ZFNs (zinc finger nucleases). More recently, TALENs (transcription activator–like effector nucleases) provide another alternative. In both of these cases, though, the resultant DNA binding protein is rather cumbersome to create and suffers from problems such as off-target binding. In practical terms, this means that it is difficult and expensive to create a new zinc finger or TALEN to target a specific DNA location for *in situ* genome editing.

In late 2012, however, the first paper was published on adapting for gene editing a defense mechanism that exists in *Streptococcus pyogenes* called CRISPR (Clustered Repetitively Interspaced Short Palindromic Repeats).[2] In combination with a protein called cas9, CRISPR binds and cuts DNA at a specific location. Unlike restriction enzymes discussed thus far, however, the exact sequence at which DNA is bound and cut is determined by a short sequence of RNA. This means that no complex protein engineering is required to target a new specific section of DNA, just a short section of RNA. This RNA is most commonly provided as DNA in a plasmid, allowing the cell machinery itself to create the guide RNA to target the CRISPR-cas9 complex appropriately. It is functionally a programmable restriction enzyme! In the year since the first publication of this approach, the CRISPR-cas9

2. Jinek, M. et al. "A Programmable Dual-RNA–Guided DNA Endonuclease in Adaptive Bacterial Immunity." *Science* 337 (2012): 816–821.

system has been used to cheaply and effectively edit DNA in humans,[3] mice,[4] zebrafish,[5] yeast,[6] bacteria,[7] and vascular plants.[8] There is reason to believe that it is a fully generalizable system for *in situ* genome editing. It has even been used already to make germline edits in mice, which were successfully passed down to subsequent generations.[9]

Although a detailed discussion of the mechanism of action of CRISPR-cas9 systems is beyond this quick introduction,[10] it is worth briefly noting the highlights of the system to better understand its strengths and limitations. First, there is a particular DNA sequence needed, known as the PAM sequence (proto-spacer associated motif), which corresponds to the particular nucleotide sequence NGG. As an indication of the frequency of this motif, there are approximately 642,000 sites on the yeast genome where the CRISPR complex can bind with maximal efficiency. Some elements of the system can be manipulated to bind with reduced efficiency if an exact PAM sequence is not located near your target of interest, so most genes can be targeted with some degree of efficiency.[11] Next, the guide and spacer RNAs have some length requirements to ensure optimal specificity. But beyond these bioinformatic tasks, the production of a highly specific CRISPR-cas9 complex consists merely of the synthesis of a few dozen bases of oligos and inserting the CRISPR

3. Mali, P. et al. "RNA-Guided Human Genome Engineering via Cas9." *Science* 339 (2013): 823–826.

4. Wang, H. et al. "One-Step Generation of Mice Carrying Mutations in Multiple Genes by CRISPR/Cas-Mediated Genome Engineering." *Cell* 153 (2013): 910–918.

5. Hwang, W. Y. et al. "Efficient genome editing in zebrafish using a CRISPR-Cas system." *Nature Biotechnology* 31 (2013): 227–229.

6. DiCarlo, J. E. et al. "Genome engineering in Saccharomyces cerevisiae using CRISPR-Cas systems." *Nucleic Acids Research* 41 (2013): 4336–4343.

7. Jiang, W. et al. "RNA-guided editing of bacterial genomes using CRISPR-Cas systems." *Nature Biotechnology* 31 (2013): 233–239.

8. Feng, Z. et al. "Efficient genome editing in plants using a CRISPR/Cas system." *Cell Research* (2013). doi:10.1038/cr.2013.114

9. Wang, H. et al. "One-Step Generation of Mice Carrying Mutations in Multiple Genes by CRISPR/Cas-Mediated Genome Engineering." *Cell* 153 (2013): 910–918.

10. Barrangou, R. "RNA-mediated programmable DNA cleavage." *Nature Biotech* 30 (2012: 836–838.

11. DiCarlo, J. E. et al. "Genome engineering in Saccharomyces cerevisiae using CRISPR-Cas systems." *Nucleic Acids Research* 41 (2013): 4336–4343.

and cas9 plasmids into your target of interest, a far cry from the complexity of previous approaches.

In all the cases discussed here (zinc finger nucleases, TALENs, and CRISPR-cas9), the end result of this specific targeting is a break in the DNA at that point. Either in only a single strand[12] or in both strands. The goal of this single- or double-stranded break is to enable new genetic material to be inserted, or to disable an existing gene. The insertion of a new genetic sequence depends on a DNA repair process known as homologous recombination—essentially built-in DNA repair machinery to knit the DNA back together. If you provide compatible DNA donor sequences, a certain portion of the time they will get stitched into that specific point of breakage in the chromosomal DNA. Or, if you cut at two adjacent points in the genome and do not provide a compatible donor DNA sequence, the target will get stitched back together without the segment you want deleted.

We're entering an exciting era in genetic engineering where we have tools accessible to not only create new genetic circuits on plasmids, but also to edit the existing chromosomal DNA of a wide variety of organisms, and to do it using tools that are low cost and accessible. At BioCurious, we've begun (but not yet seen results of) CRISPR-cas9 editing of the E. coli genome. You'll hear more about those efforts in future reports. But there's a concern: the tools to work with CRISPR-cas9, and many other related technologies, are the purview of academic researchers, and their commercial use requires individually negotiated licenses. As a nonprofit institution, BioCurious is able to order materials from Addgene, a repository of research plasmids, for research and educational use of those materials at BioCurious. As an educationally focused nonprofit, BioCurious is essentially acting in the role of a research university in supervising the material transfer agreements (MTAs) that protect the intellectual property rights of the original depositors of biological material. It's a model that seems to be working well, but blurs the line between DIY biology and the support and legal structures that underpin institutional biology. Along with safety and regulatory concerns, the legal status of the essential tools and organisms that form the core of modern synthetic biology are geared toward institutions rather than individuals. It's a core challenge of DIY biology to figure out how to work with those structures in such a way that individual freedoms to explore and learn are balanced with the needs that led to the development of the institutional control structures in the first place.

---

12. Mali, P. et al. RNA-Guided Human Genome Engineering via Cas9. *Science 339*, 823–826 (2013).

Derek Jacoby spent a decade at Microsoft Research and is now a Ph.D. candidate at the University of Victoria focused on biological data analysis. He also runs a community biology lab called biospace.ca and has been a participant and mentor on several iGEM teams.

Synbiota: An Entry-Level ELN for the Budding Scientist

Oliver Medvedik

Although promises of a paperless office may not have materialized yet, a bumper crop of ELNs (electronic laboratory notebooks) have been making the rounds to ease the gathering, processing, and dissemination of data. As we are constantly being reminded, without proper management, this torrent of data will soon overwhelm the scientific community with data anarchy, grinding progress to a halt and heralding in yet another Dark Age. Perhaps not, but the efficient management of projects when you have multiple students working on a summer iGEM (international genetically engineering machines team) becomes an absolute necessity, or things can quickly spiral out of control.

ELNs range from feature-laden systems that not only help you gather data and manage projects but, with accounting and inventory features, can essentially help you to manage an entire lab, to more basic systems that focus on individual and small group projects. Though still in beta mode, Synbiota is in the latter category and proves to be a good entry-level electronic lab notebook that allows those unfamiliar with this format of data recording and sharing to hit the ground running.

It comes packaged with a simple-to-use open source DNA annotation program, GENtle 2.0 beta, to help assemble your newly designed constructs. After uploading your sequence file, annotation is performed through a series of straightforward dropdown menus. The most useful functions, such as restriction digest mapping, are already preloaded. Files can then either be exported in a variety of formats, such as FASTA, or saved within the Synbiota.ca server for your teammates to see.

Synbiota has several registration options. In the free plan, you can have one collaborator share a notebook and have an unlimited number of public projects ongoing that can be viewed by anyone accessing Synbiota, along with one private project that only you and your teammate will have access to. All options also give users unlimited use of the open source DNA annotation tool, GENtle 2.0. Higher levels of access operate on a sliding payment scale, allowing up to 15 additional collaborators and 5 private projects.

According to Katty Wu, a rising senior at Stuyvesant High School who is currently using the site during a summer research project here at our community biotech lab, Genspace:

> ...Synbiota has been a very user-friendly tool to help organize the team's lab notebook. It's simple to use and the sidebar is a great way to navigate through all our entries. It has allowed me to easily communicate a day's work at lab with my mentors and lab colleagues...The service is extremely friendly and really values my opinions and problems. It's a site which allows starting research scientists, as well as experienced ones, to cooperate on an online environment knowing that someone's always got their back.

The site is still in beta mode, so bugs are to be expected, and new features can be added, such as a calendar to keep track of projects. I've emailed these and other suggestions to Connor Dickie (*connor@synbiota.ca*), cofounder and CEO of Synbiota, Inc. Throughout this summer, our students have had very fast and positive responses from Connor and his development team, Justin Pahara and Kevin Chen.

Currently, where Synbiota stands out is in allowing groups of students, scientists, and engineers to quickly get started using an ELN for data sharing. Uploading data is made easy through a minimalist interface that breaks everything down into four categories: Notes, Experiments, Protocols, and Literature. Having mentored four IGEM teams, I can attest to the difficulties in having students share lab notes with one another when they may be operating on different schedules. Having the information online just makes so much sense. Using a system such as Synbiota also expedites uploading of content to other sites, such as in a competition, or for eventual publication. In the end, you will have to transcribe your notebook into an electronic format anyway, so this is a great way to start getting used to the process. It also readily permits mentors to check on progress. So if you're either new to the world of ELNs or if you have a boisterous team of students to manage on a cloning project, I would definitely recommend that you give Synbiota a shot. You can register for free at *http://www.synbiota.ca*.

Oliver Medvedik is currently Science Director at Genspace, located in Brooklyn, NY, and Sandholm Visiting Assistant Professor of Biology and Bioengineering at The Cooper Union, where he is also the Assistant Director of the Kanbar Center for Biomedical Engineering.

As part of his doctoral work at Harvard Medical School, he used single-celled budding yeast as a model system to map the genetic pathways that underlie the processes of aging in more complex organisms, such as humans. He received his bachelor's degree in biology from Hunter College, City University of New York. Since receiving his Ph.D. from Harvard, he has worked as a biotechnology consultant, taught molecular biology to numerous undergraduate students at Harvard University, School of Visual Arts, and The Cooper Union and has mentored four iGEM teams (international genetically engineered machines competition) from 2009–1012.

DIY BioPrinter

Patrik D'haeseleer

Bioprinting is printing with biological materials. Think of it as 3D printing, but with squishier ingredients. There's a lot of work being done at research labs and big companies like Organovo on 3D printing of human tissues for drug testing, or even of whole human organs for transplantation. But the basic underlying technologies are surprisingly accessible: it's all based on inkjet printing or 3D printing—technologies that are readily available to the DIY scientist.

Figure 8-1. The $150 DIY BioPrinter built at BioCurious

When we first opened the doors on the BioCurious lab in Sunnyvale, we wanted to pick a couple of community projects around which we could build a critical mass —ideally projects that weren't purely limited to the wetlab, so we could have newcomers walk in the door and start participating right away. Bioprinting seemed to fit the bill. Over the course of about a year, meeting once a week with a constantly changing set of participants, we actually managed to put together a rudimentary printer that could print E. coli cells onto an agar plate using an XY platform built from parts scavenged from old CD drives and an inkjet printhead. And it actually worked the first time around! We are able to print a few lines of "I <3 BioCurious" in E. coli expressing Green Fluorescent Protein.

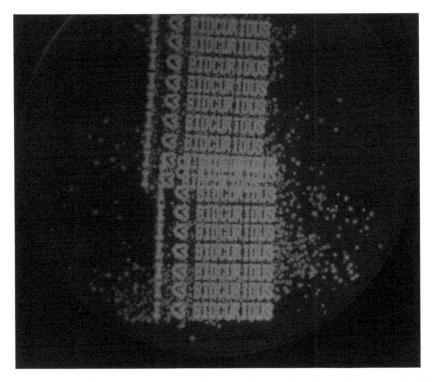

Figure 8-2. I <3 BIOCURIOUS, printed in E. coli expressing Green Fluorescent Protein on an agar plate

None of this would have been possible even five years ago. The idea to use CD drive stepper motors was borrowed from another DIYbio group, Hackteria, in Europe; the InkShield driving the print head is an Open Hardware project originally funded on Kickstarter; the InkShield and stepper motors are controlled by an Arduino; and some of the acrylic was laser cut at TechShop. None of that whole amazing infrastructure was around just a few years back.

So now that we actually had a working artefact, we needed a way to share it with the wider DIYbio community. But how do you publish your achievements in DIYbio? Some might have gone with a github repository. We decided to write a detailed description on the Instructables website, which seemed a better fit for a project how-to with lots of pictures. You can find that Instructable at *http://www.instructables.com/id/DIY-BioPrinter*. Ironically, our $150 DIY BioPrinter Instructable won two contests on the Instructables website, winning us a $1500 Macbook Pro and a $1500 UP! 3D printer—a 20× rate of return!

So what do you do with a DIY BioPrinter? Make yourself a DIY organ transplant? Not likely. Animal cells are a pain to maintain in a community lab. And human cells pose a biosafety risk, since you're growing cells without an immune system that can catch whatever nasty bugs you're carrying and spread them to other people. So rather than compete with the multimillion dollar biomedical applications that industry is pursuing, at BioCurious, we've decided to strike out into a direction where we have almost zero competition: instead of trying to print human organs, why not try...plant organs? Specifically, we'd love to print a synthetic leaf and get it to photosynthesize. The reaction we get to that idea is usually "Wow, cool!" And *that* is exactly why we're doing it. As DIY biohackers on a shoestring budget, we don't need to ask ourselves who's going to fund our research, or what commercial applications it might have—after all, there aren't too many plants desperately in need of a leaf transplant. But anything we accomplish in bioprinting with plant cells will be novel and publishable. And heck, we'll probably stumble across something with commercial possibilities, but that's not our primary motivation.

Our original BioPrinter instructable made quite a splash, with articles in Wired (*http://www.wired.com/design/2013/01/diy-bio-printer/*), MIT Technology Review (*http://www.technologyreview.com/view/511436/a-diy-bioprinter-is-born/*), Makezine (*http://blog.makezine.com/2013/04/19/how-to-diy-bioprinter/*), and a bunch of major tech blogs. We know of several groups that are building their own BioPrinter based on our example, including at least two academic labs that are planning to use it as an actual research platform. (Can you autoclave an inkjet cartridge? Only one way to find out...) The week after we published our instructable, Nicolas Lewis, the designer of the InkShield driving our inkjet print head, ran out of stock and had to start production on a new batch of InkShields.

Would we build the same BioPrinter with scavenged parts if we had to do things over again? Probably not. It was a wonderful learning opportunity for us, especially considering none of us in the BioPrinter group had any experience with bioprinting, and hardly any of us knew anything about inkjet technology, Arduinos, laser cutting, stepper motors, or how to drive stepper motors. We all taught each other what little pieces we knew, and we managed to pull off something incredibly cool. I would still recommend our current design as a learning tool that can be built on an incredibly small budget. But it definitely has its limitations: it's only an XY platform, and adding a Z stage may be problematic given the low-power CD drive stepper motors. And using an inkjet cartridge as a print head means you can only print with liquid cultures that have the consistence of inkjet ink.

Our further hardware developments since the instructable have focused on three areas:

1. Replacing the cobbled-together stepper motor drivers on our old BioPrinter with a RAMPS shield, which is the electronic of choice for the 3D printing community these days

2. Adding a bioprinting head onto the brand-new 3D printer we won (people starting from scratch may want to look at some of the very cheap 3D printer models coming out now, like the $200 MakiBox, to use as a dedicated bioprinting platform!)

3. Replacing the inkjet print head with a set of syringe pumps

Below is a DIY syringe pump we threw together: just a $10 linear stepper motor, pushing directly on the plunger of a small-diameter disposable syringe. It works great, at a cost that is a couple orders of magnitude less than professional syringe pumps.

Figure 8-3. Under construction: a $10 DIY syringe pump, capable of delivering 0.5ul/step

The syringe pumps should allow us to print using gels that maintain their shape, enabling 3D prints. Or perhaps better: print using sodium alginate and cells in one syringe and calcium chloride in a second. Where the two come in contact, the alginate will solidify, trapping the cells in a biocompatible matrix. Now imagine what crazy things you could do with cells that are engineered to express alginase under specific conditions so they can resculpt the 3D matrix that holds them in place. Add in some cell-to-cell communication, chemotaxis, and bacterial cellulose biosynthesis pathways, and you have yourself a complete toolbox for 3D pattern formation. The possibilities are endless.

This project has been an often chaotic and organically evolving collaboration between dozens of people. Too many to list here, but I'd love to thank every one of you who've contributed—you know who you are!

Patrik D'haeseleer is a bioinformatician by day, mad scientist by night. He is the community projects coordinator at BioCurious, cofounder of Counter Culture Labs, and scientific advisor of the Glowing Plant project, neither of which is in any way related or funded by his day job at the Lawrence Livermore National Laboratory and the Joint BioEnergy Institute. The views presented here are his own and do not represent those of the Glowing Plant project, BioCurious, CCL, LLNL, or JBEI.

Community Announcements

Counter Culture Labs is a new community lab for the East Bay, focused on DIY biology and citizen science. It's a place to explore, learn, work on fun projects, and tinker with biology and other sciences, and it's open to biotech professionals, scientists, and citizen scientists of all stripes. Come be part of our community of creative thinkers, hackers, and mad scientists! We are currently organizing ourselves, looking for a space to build our lab, and running meetups. Planned meetup topics include Instructables build nights, an ongoing bioinformatics series, the Science of Taste, and DIY testing of water for bacterial contamination. More info at *http://www.counterculturelabs.org/* and *http://www.meetup.com/Counter-Culture-Labs*.

We are excited to announce **Build My Lab**, a DIY lab equipment design competition hosted by Tekla Labs and Instructables. Tekla Labs (*http://www.teklabs.org/*) is a UC Berkeley–based nonprofit organization striving to empower scientists worldwide by providing online instructions for building laboratory equip-

ment from locally available materials. Submit your designs for laboratory equipment (*http://www.instructables.com/contest/buildmylab/*) from September 2nd to December 16th for your chance to win one of our fantastic prizes including a 3D printer, dremels, and more. To find out more and keep up to date on all things Tekla, like us on Facebook or follow us on Twitter @teklalabs (*https://twitter.com/teklalabs*).

Upcoming Events

Participate in our virtual journal club every other Wednesday at 5:30 p.m. PST/8:30 p.m. EST via zoom.us (*http://http://zoom.us/*). Join *diybio-idealab@googlegroups.com* for more information.

Check out these links for information on DIY biology courses and events near you:

BioCurious
Sunnyvale, California: *http://www.meetup.com/biocurious/*

Counter Culture Labs
Oakland, California: *http://www.meetup.com/Counter-Culture-Labs/*

LA Biohackers
Los Angeles, California: *http://www.biohackers.la/events*

HiveBio
Seattle, Washington: *http://hivebio.org/*. Check out the Brain Dissection class with cofounder Bergen McMurray on October 26th at 2pm.

Genspace
Brooklyn, New York: *http://genspace.org/events*. Attend the Crash Course to Biotech class taught by cofounder Ellen Jorgensen.

BUGGS
Baltimore, Maryland: *http://www.bugssonline.org/courses.html*. Check out the Build-a-BUG series to learn the basic techniques used in synthetic biology.

Brightwork Co-Research
Houston, Texas: *http://www.brightworkcoresearch.com/*

DIYbio Montreal
Montreal, Canada: *http://www.meetup.com/DIYbio-Montreal*. Check their meet-up page for upcoming events and classes.

Biospace
Victoria, Canada: *http://www.biospace.ca/classes/*

London Hackspace
London, United Kingdom: *http://wiki.london.hackspace.org.uk/view/London_Hackspace*. Attend one of the many recurring meetups, such as bio-hacking or 3D printing.

SYNBIOBETA

SynBioBeta has a few events coming up in the fall. We are running a number of introductory to synthetic biology courses in San Francisco, London, and Cambridge. We are also planning a number of advanced courses (*http://www.synbiobeta.com/courses*). In addition, the yearly "Startup Ecosystem" event will be held on Friday, November 15th in Mission Bay, San Francisco. This event offers a focal point for the syn bio startup community to meet and hear about the latest advances in the field and network with potential investors and partners. All are welcome to attend, and discounts to all our events are available to biohackers.

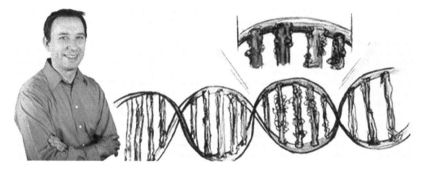